TABLE OF CONTENTS

TABLE OF FIGURES

TABLE OF TABLES

ACRONYMS

A/R	Air Refueling
ACN	Airborne Communication Node
ACTD	Advanced Concept Technology Demonstration
BAT	Brilliant Antitank (Munition)
C4ISR	Command, Control, Communications, Computers, Intelligence, Surveillance, and Reconnaissance
CBRNE	Chemical, Biological, Radiological, Nuclear, and High Yield Explosive
CMD	Cruise Missile Defense
COMINT	Communications Intelligence
CSAR	Combat Search and Rescue
DARPA	Defense Advanced Research Projects Agency
DEW	Directed Energy Weapon
DoD	Department of Defense
ELINT	Electronics Intelligence
EMP or E-bomb	Electromagnetic Pulse Bomb
FOPEN	Foliage Penetration (Radar)
FY	Fiscal Year
GPS	Global Positioning System
HPM	High Powered Microwave
IO	Information Operations
ISR	Intelligence, Surveillance and Reconnaissance
JDAM	Joint Direct Attack Munition
LOCAAS	Low Cost Autonomous Attack System
MFSP	Multifunction Signals Intelligence Payload
MOOTW	Military Operations Other Than War
PDAM	Precision Direct Attack Munitions
SAR	Synthetic Aperture Radar
SDB	Small Diameter Bomb
SEAD	Suppression of Enemy Air Defenses
SIGINT	Signals Intelligence
SMD	Small Munitions Dispenser
SSB	Small Smart Bombs
TMD	Theater Missile Defense
TUAV	Tactical UAV
UAV	Unmanned Aerial Vehicle
UCAV	Uninhabited Combat Aerial Vehicle
USAF	United States Air Force

CHAPTER 1: INTRODUCTION

Why Study Uninhabited Combat Aerial Vehicles?

The United States military has pursued the goal of an unmanned attack aircraft since before entering the First World War. Since that time, there have been periods of bust and boom for researchers in this field. Many different types of aircraft were built and used by the United States for many different missions. Though these unmanned aircraft were successful in certain missions, it was not until recently that the technology had matured enough to make uninhabited combat aerial vehicles a viable replacement for certain manned aircraft. The United States now has the technology; does it also have the desire?

With the decline of the Soviet Union and no emerging equivalent threat, there has been increasing pressure on the military to save money. President Bush and his administration have voiced a concern about spending money on merely evolutionary systems. They want to keep our military sharp but desire to transform it at the same time. The Bush administration believes the best way to accomplish this is for the military to skip a generation of weapons and their associated costs. Uninhabited combat aerial vehicles have been one attractive possibility to achieve this desire. Uninhabited combat aerial vehicles have supporters in the legislative branch as well. Senator John Warner, the ranking member of the Senate Armed Services Committee, believes the Pentagon should "aggressively develop and field unmanned combat systems in the air and on the ground."[1] He believes that by 2012, one-third of the nation's deep strike aircraft could be uninhabited combat aerial vehicles.[2]

---

[1]John A. Tirpack, "Send in the UCAVs," *Air Force Magazine*, August 2001 [article on-line]; available from http://www.afa.org/magazine/Aug2001/0801ucav html; Internet; accessed on 11 March 2003.

[2]Information concerning Bush administration is from: Donald H. Rumsfeld, "A Choice to Transform the Military" [Defenselink on-line]; available from http://www.defenselink mil/ speeches/2002/s20020516-secdef.htmlhttp://www.defenselink mil/speeches/2002/s20020516-secdef.html; Internet; accessed on 11 March 2003.

In addition to the pressure from civil authorities to transform by using uninhabited combat aerial vehicles to augment or replace manned aircraft, the senior military leaders in the United States Air Force are behind it as well. Contrary to what has sometimes been called the "white scarf" fighter pilot mentality, the Air Force Chief of Staff and other fighter pilot generals are enthusiastic about uninhabited combat aerial vehicles.[3]

The last part of this equation to consider is the American public. Are they willing to accept the notion of uninhabited combat aerial vehicles? There is certainly political pressure to minimize the risk to pilots of United States aircraft in peace or war. Procuring uninhabited combat aerial vehicles as manned warplane replacements would allow the military to continue to engage foes yet not put airmen at risk by removing pilots from dangerous or tedious missions. What about public fears of armed robot aircraft flying around? There does not seem to be any.

In autumn of 2001, the Central Intelligence Agency used an uninhabited combat aerial vehicle (a Predator unmanned aerial vehicle modified to carry Hellfire air-to-ground missiles) in operations in Afghanistan, proving its operational capability. When news of the attacks with the Predator was made public, there was no public uproar in evidence. In fact, according to an ABC report, only some Air Force pilots seem to be reacting with consternation to the new trend, out of fears for job security.[4] The American people seem to be satisfied with the reality of operational uninhabited combat aerial vehicles.

The Bush administration thinks uninhabited combat aerial vehicles save money. Senators are pushing for their acquisition. United States Air Force leaders think they will help get the job done. The public thinks they will save American lives. There is every reason to believe that

---

[3]Tirpack, "Send in the UCAVs."

[4]John McWethy, "Robo-Planes: Unmanned Aircraft Redefines How Military Wages War," [ABCNews.com on-line] available from http://abcnews.go.com/sections/wnt/DailyNews/ roboplane020501 html; Internet; accessed on 15 February 2003.

uninhabited combat aerial vehicles will be a substantial portion of the United States Air Force's arsenal by 2015.

We are currently in the midst of test and evaluation of the next generation of uninhabited combat aerial vehicles. Within the next 10-12 years, substantial numbers of these aircraft will begin to augment and replace manned aircraft in the United States Air Force inventory. Significantly, therefore they will also appear in the joint combatant commanders' toolbox. Are uninhabited combat aerial vehicles just an intriguing technology development or will they actually deliver operational utility for the joint combatant commander? This monograph seeks to answer the following question: will uninhabited combat aerial vehicles have utility for the joint combatant commander in the year 2015?

Assumptions

This paper is based on four assumptions. The first is that the basic organization of our military will remain much the same as it is now; there will still be separate services (especially a United States Air Force) and we will still have joint combatant commanders. The second is tied to the first and assumes that the basic roles of these two structures will still be the same as they are now. By this, I mean each service will still be responsible to organize, train, and equip forces to carry out its assigned responsibilities, and joint combatant commanders will still exercise these forces operationally to meet United States national objectives. The third assumption is that we will still be experiencing a moderate rate of technological advancement on par with what we have experienced over the last decade or so and not suddenly stagnate. The fourth is that regional combatant commanders will be rational in their decision making, evaluating strengths and weaknesses of each military implement in his warrior's toolbox and will then apply forces in accordance with joint doctrine.

Scope

To come to an answer for the research question, this monograph is focused on the advances made by the United States Air Force (USAF) in UCAVs. To be certain, other countries and services are pursuing these types of technologies and a combatant commander might also utilize the UCAVs of these services and nations if the need arose. However, since the United States Air Force is the major force provider for airpower and has taken the lead in producing unmanned combat aircraft, it is a bellwether for determining UCAV effectiveness by 2015. Additionally, there are families of unmanned vehicles that range from the tactical UAV to the micro-UAV and even smaller nano-UAV size ranges. These fascinating vehicles, while they may play a part in future warfare, do not easily mix with the United States Air Force's current roles and missions and are outside the scope of this paper. The date of 2015 is somewhat arbitrary, but it was the author's intention to choose a timeframe which is close enough to now (2003) to be of interest to the reader, more "real" and less "science fiction," but far enough in the future to allow technologies a chance to mature.

Methodology

In order to come to a conclusion, this paper will first delineate the history and technological progress of UCAVs. This will help the reader to understand that UCAVs have reached a stage where they are a viable operational tool for the military. Next, it will discuss the plans and predictions for United States Air Force UCAVs in the areas of doctrine, aircraft currently being tested, and UCAV weapon development. This will allow the reader to picture operational capabilities which will be available in 2015. Next, the predicted world situation will be covered along with the changes being made now to the Department of Defense to adapt to it. This provides a basis of understanding of how the military will operate in 2015. More specifically, this will explain how the joint force commander will employ his forces in 2015. Using that information as a base, this monograph will finally examine the utility of UCAVs for a joint force commander as he is making a choice on what force to employ for a desired effect by using seven criteria. The first five of these seven criteria against which the predicted characteristics and capabilities of UCAVs in 2015 will be evaluated are the required joint force capabilities discussed in chapter 4. These capabilities are: rapid global projection; forcible entry into a joint operations area; immediate and synchronized employment of scaleable joint forces tailored to conduct simultaneously a broad range of joint military operations utilizing unified action; sustainment for extended periods in an austere environment or under anti-access or area denial conditions; and continuous joint C4ISR employment and knowledge generation fusion. Additionally, potential UCAV utility will be measured against two other capabilities which go beyond the minimum required by the joint staff, those being: the ability to employ a weapon system in diplomatically sensitive situations and the capability to execute a portion of the theater missile defense mission. The measure of merit in all of these areas is whether UCAVs are currently capable of performing these in operational environments and/or whether they are predicted to do so in 2015. These criteria were chosen because it is assumed that the joint force commander will be employing forces in accord with the joint doctrine and vision set out for 2015 that is discussed in chapter 4. It is not the intent of this monograph to assert that UCAVs will be better than current manned aircraft, only that they will have the capabilities needed to be useful to joint force commanders by 2015. By proving that the UCAV will posses not just a few, but all of these capabilities, it will show the UCAV's potential utility for the joint force commander.

Structure

This analysis of the impact of UCAVs on regional combat commanders consists of five parts. Chapter 2 sets the stage for the discussion by laying out the history of unmanned aerial vehicles. In order to present an understanding of capabilities available by 2015, chapter 3 covers current United States Air Force developmental trends in doctrine, airframe and weapon technology, including acquisition timelines. Following that, in chapter 4 the author builds a composite picture of what the world will look like and how the United States military will act in the year 2015. Chapter 5 will contain a discussion on how UCAVs meet the criteria of required joint force capabilities and will examine additional benefits of employment of UCAVs for a joint force commander. Finally, conclusions and recommendations will be presented in chapter 6.

Definitions

It is important to first define the term uninhabited combat aerial vehicle (UCAV). Put simply, this is a relatively new term for an old concept: an unmanned aircraft used for combat. The first examples were called unmanned aircraft; they had no weapons onboard and since they had no people on board either, this made perfect sense. Once armed for military purposes, they were called glide bombs, flying bombs, guided bombs or aerial torpedoes, and justly so. They were on a one-way mission once launched. Eventually rockets and missiles were developed, but these essentially refer to powered warheads with control surfaces and were also one-way tickets to destruction. The drone was the first real reusable unmanned aircraft and some were used in combat, but this word has fallen out of use, except in referring to unmanned aircraft used as targets. More recently, the terms remotely piloted vehicle (RPV), then unmanned aerial vehicle (UAV) were used for aircraft in military roles. The connotation is that these aircraft are reusable and controlled by humans. You will find them at the terminus of what the author has termed the aerospace family tree depicted in figure 1.

Figure 1 is taken from a book written in 1987 and while it provides a good representation of the family of aerial vehicles, it does not include the newest progeny, the uninhabited combat aerial vehicle. Clearly, the word "combat" is a distinguishing descriptor in the phrase that plainly tells the reader that these are meant for waging war. The uninhabited combat aerial vehicle, which may operate by remote or automatic control, blurs the area on the family tree between drone and RPV. As for using the word uninhabited verses unmanned or remotely piloted, the answer might be found in connotations. A remotely piloted vehicle connotes that a human is

6

telling it what to do . . . all the time. As this paper will show, with current UCAVs that is no longer the case. Why not "unmanned" combat aerial vehicle? "Uninhabited" may simply be a gender-neutral replacement for unmanned. Or it may be that the word "unmanned" carries a negative connotation that humans have been taken out of the decisions. In the extreme this may lead to fear of flying killer robots. The term "uninhabited" does not say anything about removing a man, it just says the man lives somewhere else; the man is still in the decision loop, somewhere.

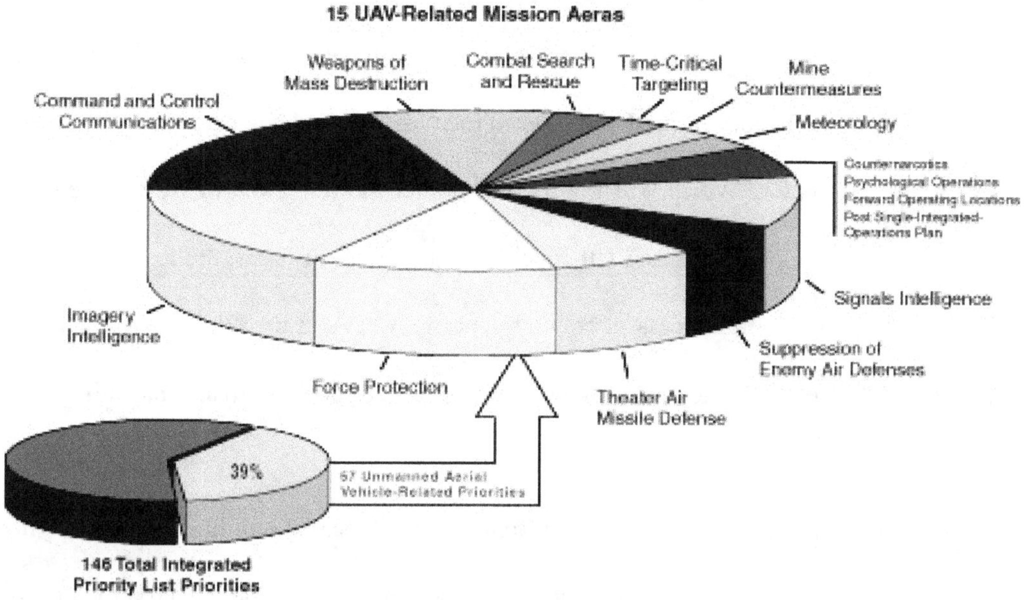

Figure 1. UAV Missions
www.airpower.maxwell.af.mil

7

CHAPTER 2: HISTORY OF UCAVS

UCAVs have generally matured along side the development of manned aircraft. Before and during World War One, France, Germany, the United Kingdom and the United States all conceived of or tried to develop unmanned aircraft. Through most of the history of UCAVs, they were never quite successful enough to recommend the redirection of critical resources to produce them in any large numbers. Many of the early attempts were simply flying bombs and while they do not meet the definition of a UCAV, they are definitely the predecessor of UCAVs. This chapter will cover the history of the UCAV up to today as follows: the early attempts to create UCAVs in the days before World War I; a rekindled interest during World War II; their increased utilization during the Cold War and Vietnam before falling out of favor; the resurgence in interest in the 1990s; and finally the successful employment of UCAVs in the Global War on Terrorism. It is a history of a weapon which has generally held out more promise than it has utility, perhaps until now.

In France, an artillery officer named René Lorin advocated building a type of flying bomb before the First World War. His proposed flying bomb was to be powered by a ram-jet or a pulse-jet, stabilized by a gyroscope and a barometer and controlled by way of radio signals from another manned aircraft. It was never built, but the concept is remarkably similar to the German V-1 flying bomb employed during World War Two. [1]

In Germany, they were working during the First World War on practical employment of glide bombs as early as October 1915. These flying bombs were essentially gliders carrying explosives. They were released from another aircraft and controlled by way of wires that unwound from a spool in the glider as it glided away from the mother ship. The control surfaces were manipulated via servomotors, which have become standard today on radio-controlled model aircraft and smaller UAVs. Batteries were first used to power these servomotors, but eventually the Germans developed airflow-driven generators instead. While the experimental aircraft eventually were able to glide about five miles, and reached a weight of over 2,000 pounds, they were not employed operationally before the armistice ended the war.[2]

In the United Kingdom, three organizations (Sopwith, De Haviland, and Royal Aircraft Factory at Farnborough) built prototype pilot-less airplanes using a new lightweight 35 horsepower engine. The others were abandoned before they flew, but the Royal Aircraft Factory built six of the machines and tested three of them. These tests involved launching the aircraft up a ramp into the air, but all stalled and crashed. With the end of the First World War the following year, British government interest in (and funding for) remotely piloted airplanes withered.[3]

Starting in 1916 aviation experts in the United States pursued what were eventually called flying bombs. Peter Hewitt, working with Elmer Sperry of the Sperry Gyroscope Company developed a workable automatic control system and installed it in a Curtiss flying boat. As the United States entered the First World War in 1917, the United States government funded further experimentation with the idea of developing a flying bomb. Five Curtiss N-9 Seaplanes made 100 flights that year. These flights consisted of the aircraft flying a pre-programmed route to an objective, while being monitored by an on-board pilot who took over manual control of the plane as it reached its destination. The United States Navy ordered five aircraft for use as flying bombs. Unfortunately the first three of these aircraft, purpose-built for being launched by catapult, crashed shortly after takeoff. As it turns out, when they put a pilot back in these aircraft, they found that the flight characteristics were significantly different from the Curtiss N-9 planes

---

[1] Sir Michael Armitage, *Air Power: Aircraft, Weapons Systems and Technology Series,* vol. 3, *Unmanned Aircraft* (London: Brassey's Defence Publishers, 1987), 1.

[2] Ibid.

[3] Ibid.

used to develop the flight control system. Further refinement experiments continued, but crashes claimed the remaining aircraft and the flying bomb program was abandoned.[4]

Concurrent with the United States Navy's program to field a flying bomb, the United States Army was also pursuing its own unmanned attack aircraft. The contract was let to Charles Kettering to develop an unmanned aircraft. He produced an unmanned biplane with the name of the *Liberty Eagle*, but it was subsequently known as the *Kettering Bug*. In January 1918, the United States Army ordered twenty-five of these aircraft to be built, six months before the first unmanned test flights. The test flights were a mixture of successes and crashes, but the Army was satisfied enough to order seventy-five more. Eventually, because of the slow process of working out the problems and the end of the war, the project came to a close with only twenty of the seventy-five Kettering Bugs ordered being built.[5] In this era of early probing of the edges of the flight envelope, manned flight all too often went beyond the edge of this envelope; it found the tasks too great for simple automated mechanical controls to handle.

During the years after the First World War, defense spending decreased in every nation that had previously been experimenting with unmanned aircraft. The United States and the United Kingdom both continued some small-scale research into gyro- and radio-controlled aircraft (to be filled with bombs or explosives). For the most part, research on employing unmanned aircraft as an offensive weapon sputtered along in the interwar years.[6]

Although aviation was in its infancy during that time, the varied used of the aircraft in the First World War opened people's eyes to the potential of aerial attack. This led to developments in ways to counter the air threat from the ground. Even in the postwar drawdowns, antiaircraft guns, and gunnery practice were considered to be something worthy of spending money on. It

---

[4]Ibid., 2.

[5]bid., 2-3.

[6]Ibid., 3-6.

was in this area that the proponents of unmanned aircraft found justification to keep pursuing the concept by providing unmanned gunnery drones. After all, what pilot wanted to fly an aircraft over anti-aircraft guns so they could practice shooting at him? In the United States from 1921-1925 unmanned aircraft were used as gunnery targets before budget concerns snuffed the programs out. The United Kingdom, however kept the unmanned target aircraft programs alive from the 1920s through to 1943. One British model of gunnery drone, named the Queen Bee, had a final production tally of 420 built from 1934 through 1943.[7]

Advances During WW II

As World War II kicked off, no nation had successfully fielded an unmanned attack aircraft. The Germans were the first to do so; they fielded the V-1 rocket, also known as the "buzz bomb," in 1944. As previously mentioned, the V-1's design was remarkably close to that proposed by René Lorin before the First World War. It utilized a pulsing ramjet engine, a aneroid barometer set to hold it at about 1,000 feet, a magnetic compass with a set heading, a gyroscope for stabilization, and a propeller-driven device to measure air miles traveled, so it could command a dive into the terrain.[8]

During the war, the United States found an enterprising way to attack highly defended targets and dispose of B-17 Flying Fortresses that had flown out their useful life span. They first stripped the bombers of all useful equipment, loaded them up with 20,000 pounds of explosives and modified the B-17 to be piloted via an open cockpit on the top of the fuselage. These were now designated the QB-17 and the program name was Aphrodite. An attack package consisted of the bomber(s), a mother ship (controlling the QB-17 by radio) and escort fighters. Once airborne and safely controlled by the mother ship, a pilot and a technician on the QB-17 would set up the

---

[7]Ibid., 2, 6.

[8]Ibid., 8-10.

radio equipment and the fuse for the explosives, then parachute out while the QB-17 was directed on toward its target. There were other aircraft like the United States Navy PB4Y, which were also used in a similar manner. All of these aircraft (most probably due to lack of precise terminal guidance) were not very effective and were vulnerable to enemy air defense and the weather. The need for this type of aircraft was not pressing in late 1944 and the hoped-for operational gains were not being realized, so Aphrodite and its sibling programs were cancelled.[9]

Both the Allies and the Axis used technological advancements gleaned from unmanned aircraft research to produce guided bombs of various sorts. These bombs were radio controlled, some radar guided, and some were even television guided. Guided bombs and missiles while having some limited successes were not produced in large numbers, nor widely used in World War Two.[10] In any case, they represent a branch of the aircraft family tree that has produced precision-guided munitions, and an expanded view of their history lies outside the scope of this paper.

Germany was the only nation to have a large-scale unmanned attack aircraft effort with their V-1 and V-2 programs. These programs gave the Germans an opportunity to keep the attacks going on England, after losing the Battle of Britain, and still minimize the loss of valuable pilots and aircraft needed on the Eastern Front with the Soviets. To sum up, the V-1 was fairly effective, much more so than any of the Allied attempts, but was essentially a crudely guided cruise missile. Along that branch of the aircraft family tree is where one can find the progeny of the V-1. As for the V-2, it was the sire of the branch yielding the modern intercontinental ballistic missile. All of these systems added to the body of knowledge needed to produce the

---

[9]Ibid., 30-32.

[10]Ibid., 19-30.

modern UCAV in various ways, but were not UCAVs. A very recognizable predecessor of the UCAV will be covered next. [11]

Cold War Progress

In 1954, Ryan Aeronautical Company was awarded a contract by the United States Department of Defense to produce the Firebee target drone to simulate high-altitude enemy aircraft. Ryan's Firebee was produced as the Q-2C for the United States Air Force, the XM-21 for the Army, and the KDA series for the Navy. The Firebee was a ground- or air-launched drone that was remotely piloted and was referred to as a remotely piloted vehicle (RPV). Reusable, it was recovered by a two-stage parachute that was deployed if it was hit, had engine problems, or by radio command. [12]

Ryan Aeronautical found that because of the small size of the RPV, radar signature augmentation was needed to simulate enemy aircraft. Even with modifications, the Firebees were still hard to hit. This led Ryan to propose a reconnaissance version to the United States military. While the drones had performed well for all of the services, only the United States Air Force Special Reconnaissance Office was interested in acquiring a reconnaissance variant when Ryan offered one in 1960. [13]

Days after Ryan representatives' first meeting with United States Air Force and Department of Defense officials about producing a reconnaissance version, a U-2 reconnaissance plane piloted by Francis Gary Powers was shot down over the Soviet Union. The Soviets paraded Powers in front of the news media and used him as leverage on the Eisenhower administration. Just weeks later, in July of 1960, an RB-47 (manned) reconnaissance plane was also shot down

---

[11] Ibid., 7-33.

[12] Robert Schwanhausser, "Unmanned Aerial Vehicles: From WWI Through Today," *1997-98 International Guide for Unmanned Vehicles* (New York: Aviation Week & Space Technology; Washington, DC: Association for Unmanned Vehicle Systems International, 1997), 10.

[13] Ibid., 10-12.

by the Soviet Union over the North Sea. As a result, interest in RPVs increased. Ryan's modified Q2-C was now capable of high-altitude, long-distance flight at near trans-sonic speeds. This new RPV (named Red Wagon) beat out a Boeing competitor called Blue Scooter, and Ryan was given a contract to study the feasibility of fitting the Red Wagon with a U-2 camera and a new guidance system. It looked like the RPV program was about to expand in a new direction, but there was a short derailment. [14]

The day after John F. Kennedy was elected, Ryan Aeronautical Company was notified that the Red Wagon program had been cancelled; no further explanation was given. We now know that the Lockheed SR-71 Blackbird had been allocated all of the funds for developing strategic reconnaissance. Over the next two years, the United States Air Force and Ryan kept development going for some sort of RPV presence in reconnaissance. In 1962 it paid off as Ryan was contracted to provide a modified Q-2C in 90 days. Ryan came up with the Model 147 or Firefly. As the Cuban Missile crisis began in 1962 two were operational. [15]

After losing a U-2 over Cuba, the Kennedy administration found out about the two Firefly RPVs; they wanted more. This led to Ryan developing the Lightning Bug. Known internally as the Model 147-B, the Lightning Bug was a higher-flying version of the Firefly and was designated the BQM-34 by the United States Air Force. The RPV program was back on track, and this time it was an extended run of success. [16]

After extensive testing, President Lyndon Johnson approved of the use of the BQM-34 over China in August of 1964. After months of successful missions, the Chinese Air Force eventually shot a Lightning Bug down and put it on display in Peking. However, since there was

---

[14]Ibid.

[15]Ibid.

[16]Ibid., 14.

no pilot on board, the international news media did not spend much time on it and it caused no international uproar.[17]

Later in 1964 the United States Air Force deployed Lightning Bugs to Vietnam. The Ryan BQM-34 was used from low level (below 500 feet) to high level (over 60,000 feet) missions for purposes of photo reconnaissance, leaflet drops, electronics and signals intelligence collection, and laying of chaff corridors to confuse enemy radar to aid manned strike packages. This was accomplished by twenty-eight different variants of the BQM-34 that were so effective they saw service for more than ten years in theater. In that time, the recovery rates for the BQM-34s were around 84 percent with 2,870 of 3,435 sorties recovered. Teledyne Ryan did experiments on the BQM-34s in 1971 and 1972 to mount Maverick missiles or electro-optically guided bombs for use in suppression of enemy air defenses (known as SEAD). None of these "killer BQM-34s" were ever used operationally. The RPV had certainly earned its stripes as a specialized platform in Vietnam, but its combat program ended with the end of United States involvement in that conflict.[18]

After the end of the Vietnam War, the United States relinquished its lead in the area of RPVs in the 1970s and 1980s. The Israelis were interested in acquiring and developing RPVs for reconnaissance and decoy missions. The Israelis eventually developed a low-cost tactical reconnaissance RPV that caught the attention of the United States military.[19]

The United States finally fielded their next operational RPV in the Gulf War in 1991, but now they were called UAVs (for unmanned aerial vehicle). The Pioneer UAV was purchased by

---

[17]Ibid.

[18]Bruce W. Carmichael et al., "Strikestar 2025" [article on-line]; available from http://www. maxwell.af mil/au/2025/volume3/chap13/vol3ch13.pdfhttp://www.maxwell.af.mil/au/2025/volume3/chap13/vol3ch13.pdf; Internet; accessed on 17 February 2003, 4.

[19]Steven J. Zaloga, "Reconnaissance Role Drives Growing Interest in UAVs," *1997-98 International Guide for Unmanned Vehicles*, (New York: Aviation Week & Space Technology; Washington, D.C.: Association for Unmanned Vehicle Systems International, 1997), 16.

the United States Navy and United States Army and used in Southwest Asia to help prosecute

Operation Desert Storm. Six systems were deployed to the Gulf War and flew 330 sorties and

more than 1,000 flight hours. Because of their ability to provide relatively low-cost targeting,

reconnaissance and battle damage assessment at no risk to pilots, the Army and Navy have used

tactical UAVs ever since.[20]

Current UAVs

Currently, the United States Air Force has two operational UAVs in the inventory. The

first is the Predator, which is a medium-altitude endurance UAV capable of more than 40 hours

of endurance. This gives it the capability to loiter over an area twenty-four hours over 500 miles

from its base. The Predator is capable of 110 knots per hour at cruise speed, 75 knots per hour at

loiter speed, is equipped with either an electro-optical and infrared sensor or synthetic aperture

radar and can carry a payload of 450 pounds. In the spring of 2001, it put that payload capability

to offensive use and became a UCAV when it successfully fired a Hellfire missile at a tank,

killing it. In November of 2002, a Hellfire-armed Predator was used to attack and kill terrorists

riding in a vehicle in Yemen. It has mostly been used to collect and transmit near real-time video

via satellite uplink or line of sight transmission for battle damage assessment and intelligence

gathering. The cost per Predator is approximately $3.2 million per aircraft.[21]

The second UAV currently in service is the Global Hawk. This reconnaissance aircraft

started life as an advanced concept technology demonstrator (ACTD); this is a fast track program

to explore new technologies and develop them quickly but in small numbers. If they work out,

---

[20]Carmichael, "Strikstar 2025," 6.

[21]General data on the Predator is from Carmichael, 6-8. The information on Predator's Hellfire test is from John A. Tirpack, "Send in the UCAVs." The story about the attack in Yemen is from Alex Belida, "Yemen al-Qaida Attack Appears to be Work of Unmanned CIA Plane," *SpaceDaily.com,* November 2002 [article on-line]; available from http://www.spacedaily.com/news/uav-02zm.html; Internet; accessed on February 2003.

they can be acquired in larger numbers and new capabilities built into the next revision (this is called spiral development). The Global Hawk is a high-altitude endurance UAV capable of more than 40 hours of endurance. This gives it the capability to loiter over an area 24 hours over 3,000 miles from its base.[22] The Global Hawk is designed to cruise at over 340 knots per hour and an altitude of over 65,000 feet. It simultaneously carries electro-optical and infrared sensors and synthetic aperture radar and is designed to perform high-resolution reconnaissance of a 40,000 square nautical mile area in twenty-four hours. The cost per Global Hawk is approximately $16 to $20 million per aircraft. On 23 April 2001, a Global Hawk took off from California, flew to and landed in Australia autonomously; this marked the first nonstop crossing of the Pacific Ocean by an autonomous aircraft. No longer just a test, it has seen operational use in the War on Terror in the intelligence, surveillance, and reconnaissance (ISR) roles over Afghanistan.[23]

There was another UAV in development called Darkstar. It was a low-observable high-altitude endurance UAV originally intended for reconnaissance purposes in situations where stealth was desired. It was designed to perform reconnaissance at over 45,000 feet at subsonic speeds in high threat environments and all weather on missions of over eight hours in length. Although it crashed on its second flight in 1996, it appeared to be making successful progress. However, in 1999 the Darkstar ACTD program was cancelled. According to a Congressional Budget Office document on UAVs, it was cancelled due to mission overlap with the Global Hawk and the low emphasis on survivable reconnaissance UAVs in a conflict. However, other UCAVs

---

[22]Ibid., 8.

[23]General data on the Predator is from Carmichael, 8-9. Cost data is from Sue Baker, "Global Hawk Program Enters Initial Acquisition," *Air Force News*, 22 March 2001 [article on-line] available from http://www.af.mil/news/Mar2001/n20010322_0402.shtml; Internet; accessed on 17 February 2003. Oceanic crossing story is from "UAV Makes Record-Breaking Transpacific Flight," Jane's.com, 27 April 2001 [article on-line] available from http://www.janes.com/aerospace/military/news/misc/globalhawk010427_1_n.shtmlhttp://www.janes.com/aerospace/military/news/misc/globalhawk010427_1_n.shtml; Internet; accessed on 17 February 2003. Information on Global Hawk now being operational is from Sue Baker, "UAV Provide Warfighters a View," *Air Force News*, 12 October 2002 [article on-line] available from http://www.af.mil/news/Dec2002/121002203.shtml; Internet; accessed on 17 February 2003.

are being tested now (the X-45, which will be covered in the next chapter) which incorporate stealth and their development may have also contributed to the death of the Darkstar program.[24]

Both the Predator and Global Hawk UAVs have been used extensively in recent conflicts for C4ISR missions. In October of 2002, the Predator UAV program logged its 50,000th flight hour. It has served as a reconnaissance platform for joint commanders in the following military operations: Operation Provide Promise (1995); Operation Joint Endeavor (1996); Operation Joint Guard (1997); Operation Joint Forge (1998); Operation Southern Watch (1999); Operation Allied Force (1999); and Operation Enduring Freedom (2001). The Predator is also being used to transmit C4ISR information directly to the AC-130 Spectre gunship for time-sensitive targeting information. According to U.S. Army Brig. Gen. John F. Kimmons, director of intelligence for U.S. Central Command, the Global Hawk's high-altitude and long endurance capabilities have been critical to ongoing CENTCOM operations in Operation Enduring Freedom. As of October 2002, Global Hawk had provided more than 15,000 images during 50 combat missions over 1,000 flight hours for the operation.[25]

This chapter has revealed how long the militaries of the world have been pursuing a pilotless aircraft that is capable of reliably conducting dangerous missions well. As was shown, those that did meet with success were highly specialized and unable to accomplish a range of missions that would recommend large-scale integration into the United States military. Based on recent developments and operational successes, they may have finally reached the point where they will fulfill their promise. As evidenced in the previous paragraph, it is certainly evident that

---

[24]General data on the Predator is from Carmichael, 9. Information on the termination of Darkstar is from Congressional Budget Office, "Options for Enhancing the Department of Defense's Unmanned Aerial Vehicle Programs" [article on-line]; available from http://www.fas.org/man/congress/1998/cbo-uav htm; Internet; accessed on 17 February 2003.

[25]The data on Predator use in current operations is from Cyndi Wegerbauer, "Predator UAV Marks 50,000 Flight Hours," 30 October 2002, [article on-line], available from http://www.ga.com/news/50000_flight html, Internet; accessed on 1 April 2003. Information on the Global Hawk's use in current operations is from Baker, "UAV Provide Warfighters a View."

reconnaissance UAVs have met the measure of merit for one of the criteria; that of continuous joint C4ISR employment and knowledge generation fusion by providing utility for joint combatant commanders in current operations.[26] As discussed in the next chapter, the United States military is increasing operational capabilities of UCAVs as well.

---

[26] Robert K. Ackerman, "Operation Enduring Freedom Redefines Warfare," *AFCEA International; The Fight for Freedom* [article on-line]; available from http://www.afcea.org/signal/tribute/EnduringFreedom htm; Internet; accessed on 22 March 2003.

CHAPTER 3: DOCTRINE AND ENABLING TECHNOLOGY

The United States Air Force makes it clear that individual roles and missions are not inextricably linked to a particular platform; the best aircraft for the mission should be assigned to attain the effect desired. For example, a B-52 bomber can provide close air support by attacking enemy troops in contact with our own, counterair by cratering an enemy runway with bombs, countersea by attacking shipping, or the mission for which it was designed many years ago, strategic attack. However, it is generally true that development of aircraft is done with the end use (capability) in mind. It is in this vein that this chapter will unfold. In order to see what the UCAV may be capable of in 2015, this chapter will describe the roles and missions that are being discussed, those being actively pursued, the aircraft and the weapons being developed for deployment on UCAVs.

Roles and Missions

In 1995, Chief of Staff of the United States Air Force, General Ronald Fogleman commissioned a study on the future called *2025*. One result of this focus on the future was a paper called "Strikestar 2025." In this paper a new doctrinal concept was discussed called aerial occupation. This new mission would be carried out by a UCAV capable of staying on station over enemy territory for extended periods, sensing, and striking multiple times until relieved by another. In this way, airpower could become an omnipresent power, not a fleeting one.

Similarly, Major George D. Kramlinger, a student at the United States Air Force School of Advanced Airpower Studies, has done a monograph examining the employment of (manned) aircraft in a peace enforcement role. In it, he said his study of Operations Provide Comfort and Deny Flight prove that aircraft can employ an airpower-centered approach (with supporting ground elements) called Sustained Coercive Air Presence (SCAP); this approach can impose a cease-fire between belligerents, create an environment for short-term ground operations, and maintain long-term stability. His caveat was that in order to coerce, the parties must have targets which they hold dear. Kramlinger explained that the most effective coercion is to threaten the parties' operational level capabilities, not their tactical level. Kramlinger's proposal does not consider the use of the role of aerial occupation by UCAVs; however, aerial occupation only enhances it.[1] These concepts reflect some of the thinking about what may be possible; what is the Air Force pursuing?

In 1996, the Chief of Staff of the United States Air Force directed a study called "UAV Technologies and Combat Operations" to study the current capabilities of UAVs and those they thought might be required of UAVs and UCAVs in the future. A scientific advisory board made up of retired officers and scientists accomplished this. In any mission the UCAV would perform, the board considered these criteria: the scenarios to be encountered, the missions and tasks, the alternatives, the relative risks, the relative costs of the tasks, and the maturity of the technologies. This board of experts delineated 22 missions or tasks (see table 1) they considered the UAV capable of performing. They recommended the Air Force focus in the near term on the first nine missions (the left column in the table) because they were high-interest missions to Air Force leaders, of utility to joint commanders, and would require sets of capabilities and technologies similar to the remainder which could later be transferred to the rest of the missions on the list.[2] These missions were to help set a course for the United States Air Force to follow in research, development and procurement of UAVs/UCAVs. The following section will lay out the progress of the United States Air Force toward these missions.

---

[1]George D. Kramlinger, "Sustained Coercive Presence; Provide Comfort, Deny Flight, and the Future of Airpower in Peace Enforcement," *Air University Press,* February 2001 [article on-line]; available from http://www.maxwell.af mil/au/aul/aupress/SAAS_Theses/Kramlinger/kramlinger.pdf; Internet; accessed on 17 February 2003.

[2]Secretary of the Air Force, Public Affairs, "UAV Technologies and Combat Operations" [article on-line]; available from http://fas.org/man/dod-101/sys/ac/docs/ucav96/chap3.pdf; Internet; accessed on 17 February 2003, 3-1 – 3-2.

Table 1. Missions or Tasks That UCAVs May Perform	
Higher Priority	**Lower Priority**
· Counter Weapons of Mass Destruction · Theater Missile Defense -- Ballistic/Cruise Missiles · Fixed Target Attack · Moving Target Attack · Jamming · Suppression of Enemy Air Defenses · Intelligence, Surveillance and Reconnaissance · Communications/Navigation Support · Air-to-Air	· Base Defense · Strategic Attack · Space Control · Special Operations · Area Denial · Decontamination & Defoliant Dispensing · Combat Search and Rescue · Trans/Post SIOP Missions · Refueling Tanker · Cargo Transport · GPS Augmentation · Information Warfare · Humanitarian Assistance

Of these higher priority roles recommended by the board, UAVs are currently being used extensively for intelligence, surveillance, and reconnaissance. According to the Department of Defense (DoD) *Unmanned Aerial Vehicles Roadmap, 2000-2025*, this role will continue in the foreseeable future and also lends itself to expanding to GPS and communications augmentation (called communications/navigation support in table 1) with operational capability by fiscal year (FY) 2006. The DoD is pursuing technology that will enable UAVs to conduct ISR in any weather and also "see" through dense foliage. This of course will aid ISR, as well as other roles.[3]

One of these higher priority roles that will be aided by such advances in ISR capabilities is theater missile defense against ballistic or cruise missiles. A UCAV combining a long loiter time and a directed energy weapon (DEW), may soon be used in roles to augment manned aircraft and space systems in theater missile defense. The rapid attack, assess and reattack

---

[3]Department of Defense, *Unmanned Aerial Vehicles Roadmap, 2000-2025* (Washington, DC: April 2001), A-1–A-24.

capabilities of a DEW and advanced onboard radar would allow a UCAV to detect a launch and possibly defeat a missile in the boost phase.

Similarly, the UCAV will be useful in the role of air-to-air combat. The concept is for UAVs to capitalize on their advantages over manned aircraft in an air-to-air scenario; they have greater tolerance for high-speed, high-"g" maneuvering to defeat other aircraft. The UCAV does not get tired after these maneuvers or bored on patrol. They could be used in interceptor, sustained combat air patrol, or high-speed ambush missions.

Countering weapons of mass destruction is a very important and complex mission. UCAVs will play a part. In the near term, the UCAV could augment manned systems by flexible, long-loiter reconnaissance and monitoring. The United States is also pursuing remote biological and nuclear detection capability which will probably be deployed for use on Tactical UAVs (TUAV). The DoD is looking for ways to use extremely high-temperature incendiary devices to incinerate chemical and biological agents without dispersal.[4] United States Air Force UCAVs will be capable of delivering those types of weapons.

The first attack mission for which the Air Force is planning to use their new UCAV is SEAD. The initial employment for this mission may be to have the UAV act as a spotter to collect on the emissions of the air defense system and pass it along to manned strike aircraft. By 2015, the Air Force's scientific advisory board felt that the UCAV should be capable of doing both sensing and striking in an operational environment. This will give the UCAV an immediate battle damage assessment and restrike capability. According to Colonel Michael Leahy, the UCAV program director at DARPA, this capability will be operational in the 2011-2012 timeframe.[5] The radar and communications jamming role will also be part of normal SEAD

---

[4]Ibid., 31, A-

[5]Roxana Tiron, "Unmanned Bomber Prepares for Crucial Tests," *National Defense Magazine,* May 2002 [article on-line] available from http://www.nationaldefensemagazine.org/article.cfm?Id=792; Internet; accessed on 17 February 2003.

operations and will be built in to the X-45. According to the Air Force Research Lab (AFRL), the development of capabilities for SEAD will go hand in hand with precision attack capability, and the first planned operational UCAV will have this capability. This UCAV is the X-45.[6]

Aircraft Now In Development

The Air Force is currently developing airframe, weapon/payload, power plant, and controlling technologies to enable UCAVs to be effective in war. By looking at some of these advances that are public, one can get an idea of what capabilities will probably exist in the operational realm by 2015. Specifically this section will address aircraft and weapon advances.

The airframe now in development and testing as an ACTD is a UCAV called the X-45. There are currently two versions, the X-45A and the X-45B. As this paper is written, there are two X-45As and they are undergoing successful flight-testing at Edwards Air Force Base. The X-45A is nearly all-electric and is designed to fly at 0.8 mach at 40,000 feet with a current flight duration of about ninety minutes. Coming next is the X-45B, which is larger than the A model with a longer flight duration; it is being evaluated to perform the SEAD mission as its first operational assignment. Boeing Phantom Works is the contractor working with the Air Force and the Defense Advanced Research Projects Agency (DARPA) to produce this next generation, purpose-built combat aircraft. The X-45 concept calls for a stealthy UCAV capable of carrying two tons of GPS-guided ordnance or directed energy weapons deep behind enemy lines for long loiter periods with electronic warfare capabilities roughly equal to the formidable (but now retired) F-4G Wild Weasel. All of this comes with a predicted cost up to 65 percent less than future manned aircraft and 75 percent less than current manned aircraft to operate and maintain.[7]

---

[6]DARPA, "Unmanned Combat Air Vehicle Summary Fact Sheet" [article on-line]; available from http://www.globalsecurity.org/military/library/news/2002/04/fs_x-45A.pdf; Internet; accessed on 11 March 2003, 1-2.

[7]All except cost data is from: Tirpak, "Send in the UCAVs," 2-5. Cost data on X-45 is from: DARPA, "Unmanned Combat Air Vehicle Summary Fact Sheet."

The type of program the Air Force is using is called spiral development. Once testing is complete and basic design parameters have been met, DARPA and the Air Force can accelerate the acquisition of small numbers of operational UCAVs very quickly; they can add minor changes during this and save larger ones for later models or blocks. The Air Force is planning to develop them in at least three blocks.

Using spiral development, the X-45B (Block 10) could be fielded by 2005 and the Air Force is programmed to field fourteen no later than 2008. The goal is to acquire as many as sixty Block 10s. Block 10 UCAVs would attack heavily defended air defense targets with its capacity of twelve of the new 250-pound small smart bombs (SSB) that have the same destructive capacity as a current 2,000-pound bomb. This mission is currently done by the F-117 and B-2 for manned aircraft and also by cruise missiles. The Block 20 model will have reactive SEAD capabilities, meaning that it will be able to see and shoot the air defense targets. Block 30 models are the ones planned to carry directed energy weapons (HPM or lasers) and will be capable of engaging stationary or moving targets on the battlefield. [8]

Where is the pilot? There will not be one, neither in the aircraft nor behind the controls in its command center. The X-45B's human partner will be a supervising operator who will monitor his package of UCAVs (from one to five). The limited human input will be to check on the mission capable status of the UCAV and give consent on weapon release, and to add flexibility to re-role the aircraft in flight to another mission or route. [9]

To employ a UCAV using fully autonomous operation, especially with lethal weapons, may require a big shift in attitudes. Some might argue that it will not happen any time soon, regardless of the capability. George K. Muellner, a retired United States Air Force Lieutenant

---

[8]Date of fielding is from Tirpak, "Send in the UCAVs," 3. Descriptions of block acquisition from John Tirpak, "Heavyweight Contender," *Air Force Magazine*, July 2002, [article on-line]; available from http://www.afa.org/magazine/July2002/0702UCAV.pdf; Internet; accessed on 17 February 2003.

[9]Ibid.

25

General and President of Boeing Phantom Works, says otherwise. The X-45 does not require any direct human involvement (after programming) to start, taxi, takeoff, fly a mission and land; only when the aircraft identifies the target does it ask for human confirmation that it is a legitimate target so that it may strike. Soon, he says the machine will probably be trusted handle that task as well. [10]

One other interesting aspect of the X-45 is its cold storage capability. It is designed to be boxed up, with its wings removed, for up to ten years and then be unpacked and made mission-ready within thirty minutes. Six of the boxes could fit in a C-17 and be airlifted almost anywhere. The boxes are designed to keep the climate right for the UCAV and also monitor its health. Instead of squadrons of maintainers, a few people could monitor them and ready them for flight when needed. Recently, in a move to make the X-45 even faster to the fight, the Chief of Staff of the United States Air Force General John P. Jumper said that they are making some changes. The Air Force is adding air-refueling capability and additional internal fuel tanks, so that they may be deployed faster and not require any theater basing. [11]

Another aircraft that the United States is pursuing is the Helios prototype project. This one, while not being developed by the military (it is being developed through the National Air Space Administration) will have capabilities that would be very helpful in many military missions. The Helios is being developed as a solar-powered UAV capable of ultralong duration (up to six months), very high altitude flight (from 100,000 to 120,000 feet) that is capable of carrying small payloads. On 13 August 13 2001 it achieved a record altitude (the highest ever flown by a non-rocket-powered aircraft in sustained horizontal flight) of 96,863 feet. It is scheduled in 2003 to attempt to stay airborne above 50,000 feet for 96 hours. Using solar energy

---

[10]Tirpak, "Send in the UCAVs," 5.

[11]Tirpak, "Heavyweight Contender."

to power itself and send excess power to fuel cells during the day, it depends on the fuel cell to power it at night.[12]

Developing Weapons for UCAVs

The first UCAVs are programmed to be capable of employing many of the same types of weapons that current strike fighters employ. These kinetic weapons use explosives or warheads to hurl shrapnel or dense metal penetrators at high velocities to physically damage targets. UCAVs are expected to employ the joint direct attack munition (JDAM) which is a 2,000-pound glide bomb guided by global positioning system (GPS). These UCAVs will also be capable of employing numerous small-diameter bombs (SDB), also previously referred to as the small smart bomb (SSB). A UCAV weaponization program is under way now to: extend the range of the SSB; enable UCAVs to carry the low cost autonomous attack system (LOCAAS); field a small munitions dispenser (SMD); and design and field a precision direct attack munitions (PDAM) for increased target accuracy. All of these are to be operational by FY 2005. Currently, the Predator UAV is capable of firing both the Hellfire antitank missile and the AIM-9 air-to-air missile for use against aircraft. Another antitank weapon which may be carried by Air Force UCAVs is the brilliant antitank (BAT) munition. This munition was successfully tested on the United States Army's smaller Hunter tactical UAV in October of 2002, but it could be expected to be fielded on other UAVs if it is effective, as has been done with the Hellfire missile. Presumably, the other types of bombs and missiles currently in the Department of Defense inventory may, with some modification, be carried on Air Force UCAVs.[13]

---

[12]Department of Defense, *Unmanned Aerial Vehicles Roadmap, 2000-2025,* A-26; and NASA, *Helios* [Dryden Flight Research Center website on-line]; available from http://www.dfrc.nasa.gov/ Research/Erast/helios.html; Internet; accessed on 15 February 2003.

[13]Information on UCAV and planned weapons from the Boeing Company, *Unmanned Combat Air Vehicle (X-45) Factsheet* [document on-line] available from http://www.boeing.com/phantom/ucav.html; Internet; accessed on 15 February 2003. The information on the Predator UAV comes from Barry Rosenberg, "UAVs Step Into Combat Arena," [Aviationnow.com on-line] ;available from http://www.awgnet.com/shownews/01paris1/hardwr10.htm; Internet accessed on 15 February 2003.

Some of the recent weapons advances that match up well with UCAV technology are nonkinetic weapons. These nonkinetic weapons use directed energy (microwaves, electromagnetic pulses, or lasers) to defeat, disable, or destroy their targets and are planned for use by UCAVs; some will be ready sooner than others. The most promising is the high-powered microwave or HPM. Using energy generated by the UCAV's jet engine, it will have the capability to shoot thousands of shots at over 100 targets in one sortie without needing to return to base and rearm. The HPM will be capable of offensive attack missions, as well as self-defense; it may be operationally available as soon as FY 2006. Also capable of a nonkinetic kill is the electromagnetic pulse bomb (called the EMP bomb or sometimes the E-bomb). It uses an explosion to generate an electromagnetic pulse to overload electrical components in targets within the "kill zone" which contain electronics, disabling them. Any electrical components that are outside of the kill zone but wired to devices within it can be damaged as well. The EMP bomb can be fitted into the same space on the UCAV as the JDAM (using a Mark-84 form factor). It is available for use now.[14] Chemical lasers are currently being tested and evaluated for use by airborne platforms, but at this time are not being actively pursued for deployment in UCAVs because of their large size. This of course, could quickly change.

Information is also a weapon. According to the DoD dictionary, information operations are defined as: "Actions taken to affect adversary information and information systems while

Hunter UAV testing of BAT is from "UAV BATs A Few Direct Hits," *Spacedaily,* 23 October 2002 [Spacedaily.com on-line]; ,available from http://www.spacedaily.com/news/uav-02zk.html; last accessed on 15 February 2003.

[14]Information on HPM is from David A. Fulghum, "USAF Acknowledges Beam Weapon Readiness," *Aviation Week & Space Technology*, 4 October 2002 [article on-line]; available from: http://www.grn.es/electropolucio/avio41002 htm, (15 February 2003). EMP bomb source is Carlo Kopp, "The Electromagnetic Bomb a Weapon of Electrical Mass Destruction," *Air and Space Power Journal Chronicles* [article on-line]; available from http://www.airpower maxwell.af mil/airchronicles/kopp/apjemp html; Internet; accessed on 16 February 2003. EMP bomb availability information is from Michael Smith, "Saddam to be Target of Britain's 'E-bomb'," telegraph.co.uk, 26 August 2002 [article on-line]; available from: http://news.telegraph.co.uk/news/main.jhtml?xml=/news/2002/08/26/wirq26.xml; Internet; accessed on 15 February 2003.

defending one's own information and information systems." The UCAVs currently in production and testing utilize stealth technology to deny the enemy information about their presence. Additionally, the United States Air Force will seek to improve its knowledge of the enemy. As previously mentioned, the UAV is currently being used extensively to collect information. The Department of Defense has prioritized the development of radar capable of seeing "tanks under trees" with radar. This new radar is called the foliage penetration (FOPEN) radar and is part of an ATCD being jointly conducted by the United States Air Force and the United States Army; according to the *Unmanned Aerial Vehicles Roadmap, 2000-2025*, it could be operational by FY 2004. The Department of Defense is pursuing a scaleable Airborne Communication Node (ACN) payload for deployment in various UCAVs; this node will not only be able to act as a relay for friendly communications, it can also be used to gather signals intelligence (SIGINT) on the enemy and ties in to a theater information infrastructure. The ACN will be flight tested by FY 2003, but no data is available on an operational readiness date. Other options for intelligence gathering are being pursued in one package called the multifunction signals intelligence payload (MFSP); it is capable of collecting communications intelligence (COMINT) and electronics intelligence (ELINT) in the same payload and is anticipated to be fully operational by FY 2004. Another friendly information tool being developed is the airborne GPS pseudo-satellite. This will put high-powered GPS-like satellites over the battlefield. In case of GPS signal jamming, the power and proximity of these pseudo-satellites will allow their signal to "burn through" the jamming nose and provide information for GPS systems. This capability will be operationally available by FY 2006. Also available for use now by UCAVs are leaflet bomb units (LBU-30) which allow aircraft to deliver large numbers of leaflets for information operations from high altitude.[15]

[15]Definition of IO: Department of Defense, *DOD Dictionary of Military and Associated Terms*, 14 August 2002 [document on-line]; available from: http://www.dtic mil/doctrine/jel/doddict/data/i/index .html; Internet; accessed on 16 February 2003. Department of Defense, *Unmanned Aerial Vehicles*

This chapter has covered some possible doctrine for UCAVs, what roles and missions the United States Air Force is currently pursuing for their UCAVs, and the UCAV-related technology that will likely be in operational play in 2015. In summary, the operational capabilities of US Air Force UCAVs will be sufficient by 2015 to perform precision strike (and re-strike) using a range of kinetic and directed energy weapons, SEAD, and improved C4ISR integration. They will do this with anywhere from a small in-theater presence to no in-theater presence and very little human interaction. When initially fielded, UCAVs will operate in concert with other combat aircraft as part of a combat package (SEAD, pre-strike reconnaissance, strike, and post-strike reconnaissance); by 2015 a UCAV will be capable of performing the roles of the whole package by itself (all of the above and stealth for self-defense). These UCAVs will be nearly autonomous and will handle long, boring and dangerous missions effectively. In order to answer the research question, this paper will now turn to look at the expected operating environment for the United States military and the joint force commander in the coming years.

Roadmap, 2000-2025, A-13--A-24. Information on leaflet bombs: Robert Wall and David A. Fulgham, "New Tools Emerge For Info War Battle," *Aviation Week & Space Technology,* 26 February 2001 [article on-line] available from: http://www.cadre maxwell.af mil/warfarestudies/iwac/Downloads/IW270%20Reading.doc; accessed on 16 February 2003.

CHAPTER 4: FUTURE OPERATING ENVIRONMENT

What will the international environment be like in 2015? This chapter will look at the world situation as predicted by the CIA and the Joint Staff. Following that, it will discuss the operational themes and capabilities the United States military will be required to employ. Finally, it will look at how the UCAV fits into that future environment.

The World Situation

In February of 2000, the United States Central Intelligence Agency (CIA) published a forecast called *Global Trends 2015: A Dialogue About the Future With Nongovernment Experts*. In this report, the CIA experts predict an environment in which the United States will remain the world's primary superpower. As the United States pursues its national interests it will continue to face global instability caused by states with horrible internal conflicts, trans-national terrorism, inter-state conflicts, and asymmetric attacks on the United States as the sole superpower. By asymmetric attacks it means that for the most part, nations will choose to challenge us through indirect means rather than facing our strengths. These attacks will include information operations, threats to "critical infrastructure" (cyber or physical attacks on computer, transportation or energy networks), terrorism, and weapons of mass destruction.[16]

The Joint Staff has a similar view of the strategic and operational world in the future. Strategically, the Joint Staff believes that the United States will continue to have global interests and continue to protect them, the joint force battlespace will extend from the United States proper to the entire globe and include cyberspace and space, asymmetric threats against us at home and abroad will continue, the proliferation of accurate and relatively inexpensive missile technology

---

[16]Central Intelligence Agency, US National Intelligence Council, *Global Trends 2015: A Dialogue About the Future With Nongovernment Experts* (Washington, DC: 2002), 49-58.

is a growing threat abroad and at home, the threat from chemical, biological, radiological, nuclear, and high yield explosive (CBRNE) weapons will increase, operations including the military will continue to be joint, multi-agency and multinational, coalitions may be short or long term partnerships, states with internal conflicts (over ethnicity, religion, and scarce resources) will continue to cause regional instability, potential adversaries will have access to commercial sources of technology which may allow them to challenge United States superiority, these adversaries will adapt to our efforts to block or defeat these capabilities, urbanization and population will change the physical and political makeup of nations.[17]

At the operational level, the Joint Staff assumes the following: successful military ventures in the future will continue to require highly qualified personnel who are trained to excel in a joint force context; the continued potential of major theater war with overlapping regional conflicts or crises; continued compression of the levels of war (where tactical actions have strategic consequences); information operations and knowledge management will be key to joint force effectiveness; capabilities-based forces which synergistically leverage broad capabilities of existing and emerging air, land, maritime, cyber and space will be the rule; command and control will be by a common networked joint command, control, communications, computers, intelligence, surveillance, and reconnaissance (C4ISR) system that will overcome the challenges of interagency and multinational interoperability; and finally, the environment in which our forces will operate will increasingly include complex, urban and densely populated terrain. The Joint Staff has made these assumptions about the future in order to counter them, or at least prepare for them.[18]

---

[17]US Joint Staff, An Evolving Joint Perspective: US Joint Warfare and Crisis Resolution in the 21st Century (Washington, DC: US Government Printing Office, January 2003), 2-3.

[18]Ibid., 3-4.

The United States military services have been engaged since the end of the Cold War in trying to understand and prepare for a new operating environment. This environment included new emphasis on uses of the military in situations other than simply to deter an enemy from using military force and when deterrence failed, to fight and win the nation's wars. Additionally, instead of basing our military force doctrine and procurement on the predominant threat, the Department of Defense has switched to a capabilities-based model. Because in today's environment, it is not possible to know what nation or combination of nations and non-state actors may try to use military force to attack the vital interests of the United States or its allies. It is possible to predict with some confidence what capabilities any of these actors may try to employ against the United States military. Thus, this model compels the Department of Defense community to focus on how any potential adversary might choose to fight us, including asymmetric or idiosyncratic methods; it then calls on the Department of Defense to sustain and/or develop desired capabilities for dominance in all military situations as the objective for planning, procurement and doctrine. In short, the modern United States Military must be capable of countering the kind of threats that the Central Intelligence Agency and Joint Staff have said will be prevalent in the future.[19]

In recent years, all of the services have been pursuing changes in concepts, technologies, and capabilities to meet these new challenges. The President of the United States and the United States Secretary of Defense have directed the DoD to transform to be able to succeed in this new environment. The services have been, for the most part, pursuing these paths separately. In order to ensure a synergistic application of effort, the Joint Requirements Oversight Council has recently approved the release of a Joint Chiefs of Staff white paper entitled "An Evolving Joint Perspective: US Joint Warfare and Crisis Resolution In the 21st Century." This document lays

---

[19]US Department of Defense, *Quadrennial Defense Review Report, September 30, 2001* (Washington, DC: US Government Printing Office, 2001), iv, 13-14.

out some operational themes and joint team capabilities which are aimed at keeping United States military forces dominant in the future environment. These themes and capabilities define the joint environment in which UCAVs are likely to be employed in 2015 and therefore warrant closer inspection.

New Operational Themes

New operational themes were originally laid out by the Bush administration in the 2001 *Quadrennial Defense Review Report* and have since been condensed and systematically delineated in the previously mentioned Joint Staff white paper. They are meant to be integral to joint and service requirements generation and force planning processes.

The new operational themes call first for a shift in the capability of the United States to project a large force across the globe in a long period of time to the capability to project a smaller, yet more capable force across the same distance much faster. The second is for tailored combat forces that are joint and expeditionary in nature, and immediately employable once arriving in a forward location. The white paper defines an expeditionary force as "an armed force organized, trained, and equipped for rapid deployment, immediate employment in the operational area, and sustainment under austere conditions." These forces include global strike capabilities, continental United States and forward based forces, and special operating forces to augment any expeditionary or forward based combat forces already in place. The third theme is that joint maneuver and precision strike will be "at varying depths, in all weather and terrain, to deny sanctuary, attack critical vulnerabilities, or defeat the efforts of an adversary even within distant anti-access and area-denial operational environments." The fourth theme is that joint operations will be planned and executed as network-centric and effects-based expeditionary operations on a global scale. The fifth is that joint operations will have an increased emphasis on global joint C4ISR. The sixth operational theme is enhancing space operations and information operations capabilities as competencies important to our asymmetric advantage. The seventh theme is developing and deploying a missile defense capability to protect the United States homeland, friends and allies of the United States as well as forward deployed forces.

Required Joint Team Capabilities

The required operational capabilities of joint warfare and crisis resolution logically flow from these new themes. In short, these required capabilities are: timely global projection; forcible entry; immediate and synchronized employment of tailored joint forces; joint, unified action; sustainment for extended periods regardless of the environment; and focused and continuous joint C4ISR. The expanded list is shown in table 2. According to the Joint Staff, these capabilities are essential to the new vision of joint warfare.

Table 2. Joint Team Capabilities Of Joint Warfare & Crisis Resolution

• Timely global projection of a tailored joint force;
• Forcible entry into a joint operations area, when necessary;
• Immediate and synchronized employment of scaleable joint forces tailored to conduct simultaneously a broad range of joint military operations utilizing unified action;
• Sustainment for extended periods in an austere environment or under anti-access /area denial conditions; and
• Focused and continuous joint C4ISR employment and knowledge generation fusion.

Source: "An Evolving Joint Perspective: US Joint Warfare and Crisis Resolution In the 21st Century," (Washington, D.C., January 2003.)

The previous pages covered the vision of the Department of Defense for ways of thinking about and preparing for future military operations with the operational themes and the required capabilities to successfully perform their duties. The next chapter will evaluate the capabilities and characteristics of the UCAV to see if they will make the varsity team when the joint force commander goes to war in 2015.

CHAPTER 5: UCAVS, AN OPERATIONAL FIT?

In order to be a relevant combat system to the joint force commander of 2015, the UCAV must be of operational utility. Though the United States Air Force is the service that is developing the X-45 UCAVs and their follow-ons, the joint force commander will be the person who employs them. This chapter will build upon the information of the previous chapters to bring the reader to the conclusion that UCAVs will provide that utility. It evaluates the utility of the UCAV to the joint force commander by comparing what we expect UCAV capabilities will be (from chapter 3) to the desired joint force capabilities for the future (from chapter 4). The first part of this chapter will show its utility to the joint force commander by weighing the UCAV on the previously introduced criteria of joint team capabilities of joint warfare & crisis resolution: rapid global projection; forcible entry into a joint operations area; immediate and synchronized employment of scaleable joint forces tailored to conduct simultaneously a broad range of joint military operations utilizing unified action; sustainment for extended periods in an austere environment or under anti-access or area denial conditions; and continuous joint C4ISR employment and knowledge generation fusion. Finally, this chapter will also address the two other capabilities beyond these minimum requirements envisioned by the Joint Staff which are: the ability to employ a weapons system in diplomatically sensitive situations and the capability to execute a portion of the theater missile defense mission. Does the UCAV possess these capabilities sufficient to be of utility to the joint force commander?

UCAVs and the Required Joint Team Capabilities

The first capability to be required of joint force of the future is rapid global projection. There are two aspects to this capability. First, future forces (including all of their required augmentation) must be light and small enough to be rapidly deployable in an environment of constrained airlift. Second, the forces must be able to respond quickly to the deployment order, preferably without needing extended time for setting up the logistics piece to deploy it. For example, current systems may require setting up an airlift and/or an air refueling bridge, having to move special loading and unloading equipment, having to deploy or build special hangars, etc. In two ways, the UCAV is being developed to equal this challenge.

How does the UCAV meet the requirement to deploy rapidly? People are one slow part of the deployment process, and the UCAV will minimize the number of personnel required to deploy forward with it. As quick to deploy as current United States Air Force manned weapon systems are, they have to process and send operators, command and control systems and personnel, planners, ISR forces and equipment, maintainers, security, spare parts and diagnostic equipment, and leadership for this entire team, to name a few. When it is best for the UCAV to be forward based, it can deploy with anywhere from a greatly reduced number to almost none of these personnel. UCAVs are being designed with reduced maintenance requirements, and will get hands-on maintenance only when sensors indicate the need. Modular avionics will reduce the number of specialists required for actually accomplishing the maintenance work, around 70 percent fewer personnel according to one study. It gathers its own ISR, so it does not need to rely on manned reconnaissance aircraft. In addition to requiring fewer people to deploy, the UCAV (and augmentation system) is relatively small to transport. This helps to lighten the entire deployment load for the joint force commander and enables the joint team to close out the deployment phase and start operations more quickly.[1]

If the joint force commander so chooses, use of UCAVs in a theater can require no airlift sorties during the normally hectic force build up phase. As mentioned in chapter 3, the X-45 has

[1]Deployment and employment options discussed in: Keith Rogers, "Reports Say Predators Seeing Action in War," *Las Vegas Review-Journal,* 20 October 2001 [article on-line]; available from http://www.lvrj.com/lvrj_home/2001/Oct-20-Sat-2001/news/17263620.html; accessed on 24 March 2003. Information on the reduced maintenance requirements is from: Jan Walker, "Unmanned Combat Air Vehicle Operational System," Defense Advanced Research Projects Agency Fact Sheet (Arlington, VA: DARPA, April 2002), 3. The number of 70 percent reduction comes from a study quoted in: Eileen M.

a ten-year cold storage capability; it can be forward deployed to potential theaters of operation and wait for employment orders. As previously mentioned, a single C-17 can fit from four to six of the X-45 containers in one lift. In instances where it must immediately employed upon entry into a theater, it will also be able to self deploy (with or without air refueling en route depending on the distance) under new operational requirements generated by the Air Force. The command and control trailer from which multiple UCAVs can be operated will be transportable by air to theater, but it need not be there. Currently, the command trailer for the Predator UAV is C-130 deployable and can connect with the Predator by radio or satellite communication, enabling operation of the UAV anywhere in the world from the United States. In the semi-autonomous mode envisioned for future UCAVs, one operator will monitor multiple UCAVs and only be required to approve strikes or change routes based on the situation. This can be done from the United States as well, eliminating the need to deploy operators and technicians who maintain the command and control trailer and its systems. With air refueling as an option, a UCAV could stay airborne indefinitely as long as it was mechanically able.[2] All of these characteristics show the UCAV will possess a strong capability regarding the first criterion of rapid global projection in 2015.

The second capability the Joint Staff considers necessary to operate in the twenty-first century is forcible entry into a joint operations area. Seen as a "first day of the war" kind of platform, UCAVs are being developed with forcible entry capability in mind. Their stealth design (very low radar and infrared observability) is only the first of its attributes which will enable this. As stated before, its first operational mission will be SEAD; its second will be precision strike. Both of these missions are required in a forced entry situation and enable other

---

Walling, *High Power Microwaves; Strategic and Operational Implications for Warfare*, (Maxwell AFB, AL: Air University Press, May 2000), endnote 19.

forces to penetrate the area of operations more easily. The UCAV will also capitalize in another way on its small logistics footprint; it will make possible a very small deployment signature, avoiding things such as large numbers of cars gathering in the airfield parking lot, large numbers of military people telling friends and family they are deploying, an increase in people having their mail held and paper delivery suspended, etc. These things will make it hard for a foe in a far-flung corner of the world to know it is coming through surveillance of UCAV home bases. This element of surprise could add synergistically to its ability to suppress enemy air defenses and execute strike missions to enable forced-entry operations. Clearly, the UCAV will be fully capable of forcible entry requirements, which is the second criterion for utility.

The third requirement is what the joint white paper calls "immediate and synchronized employment of scaleable joint forces tailored to conduct simultaneously a broad range of joint military operations utilizing unified action."[3] This is a mouthful. It means: forces should be ready to employ immediately upon reaching a theater, not needing substantial amounts of time to build combat power; forces must be able to integrate with others in order to synchronize operations, regardless of service; forces must be scaleable enough to provide flexibility to the joint force commander in situations requiring small or large sized forces; forces must be trained and ready to employ across the spectrum of conflict. The UCAV will fit right in. Since it is an airpower weapon, the UCAV has can transit large distances quickly (as a reminder the X-45 cruises at high subsonic speeds) to employ its payload throughout the depth of the battlespace in support of joint force objectives. Additionally, it will have the flexibility to change its profile in flight so that it may stay synchronized with changes in other joint force missions. With an ability to carry out stealthy ISR, sensing and striking, immediate battle damage assessment and re-strike,

---

[2]Information on the Predator is from: Rogers, "Reports Say Predators Seeing Action In War." Information on USAF changes for UCAV from: David A. Fulgham, "UCAV Spending Spikes in Pentagon Budget Plans," *Aviation Week and Space Technology*, 17 February 2003, 56-57.

[3]US Joint Staff, 11.

the UCAV will be easily tailored to execute a range of missions; all of these coming with a relatively small logistics cost which would recommend it for inclusion in many scenarios. It is being designed to be survivable in a major theater war environment. However, its changeable payloads will make it valuable across the whole spectrum of armed conflict.[4] In low-intensity conflict, the ability of the UCAV to provide continuous on-call firepower with relatively low theater logistics requirements and (with directed energy weapons) low-collateral damage recommend it for this mission as well. In peace enforcement and peacekeeping operations, it would be able to loiter and observe targets and respond appropriately with weapons or cameras as needed. In a humanitarian relief operation, its extensive C4ISR ability would transmit much needed information to decision makers. Also, the UCAV's ability to carry a communication suite, such as the aforementioned airborne communications node or GPS relay, will enhance ground operations in all areas of the globe, especially where these services are degraded or scarce. The UCAV will thus be capable of meeting this requirement in 2015, which is the third criterion for utility to the joint force commander.

The fourth joint team capability seen as critical for twenty-first century operations is sustainment for extended periods in an austere environment or under antiaccess or area denial conditions. Part of the concern of this requirement refers to the ability to conduct a deployment to a theater quickly, even under conditions where area denial weapons (such as nuclear, biological, or chemical weapons) been employed. This paper has discussed previously that the UCAV can be employed without ever landing in theater, which would obviously be a good option in cases such as this. When required to stay in theater, because of the aforementioned relatively small logistics requirement involved in deploying and employing the UCAV, it could deploy quickly and with fewer people being exposed to the austere or hostile environment. Additionally its small logistics burden enables sustainment for extended periods in merely austere

---

[4]Jan Walker, "Unmanned Combat Aerial Vehicle Operational System," 1-3.

environments, as required by the joint staff. In situations where directed energy weapons can be used instead of kinetic weapons, the sustainment burden is lessened even further. Concerning this requirement, UCAVs in 2015 will be fully capable of sustainment in the required conditions, which is the fourth criterion for utility.

The final required joint capability is "focused and continuous joint C4ISR employment and knowledge generation fusion." As stated in chapter 2, the UAV is excelling now in helping to generate C4ISR data for joint force commanders. Even without further development of UCAVs in this area, this criterion may be considered to be met. However, continuing to increase fusion of C4ISR data into a joint common operational picture is currently is a goal of the Department of Defense. Programs now under way in UCAV-related fields such as the airborne communications node and spectral infrared remote imaging transition testbed will ensure UCAVs will continue to be a critical contributor to the joint combatant commander in 2015.[5] Plainly, the UCAV will be fully capable of the C4ISR requirements in 2015, which is the fifth criterion for utility.

Beyond Required Capabilities

Because of the predicted conditions of the future operating environment discussed in chapter 4, there are two desired weapon system capabilities (beyond the minimum required discussed previously) that may impact the joint combatant commander in 2015. These are the capability to be employed in diplomatically sensitive situations and the capability to conduct theater missile defense. The following section will address whether UCAVs possesses these desired capabilities.

The first capability which is desirable is the ability to employ a weapon in diplomatically sensitive operating environments, especially in an era where the United States has fewer overseas bases. In the early days of Operation Enduring Freedom (the military operations in Afghanistan starting in autumn 2001) the Bush administration wanted to begin military action quickly. They were hampered in their ability to get combat aircraft into the air over Afghanistan for two big reasons. First, they needed basing for combat search and rescue (CSAR) aircraft in a neighboring country, which took major diplomatic initiatives. Second, they needed approval for use of another country's airspace for use by refueling and strike aircraft. They found some countries were willing to let their country be used as a base for a small rescue force, but preferred to have it remain unpublicized. This solved only part of their problem.[6] It is one thing to allow ones country to be used to help a rescue along, but another altogether to allow your country to be used as a striking platform.

In some instances in the future we may see similar access problems where it would be preferable to fly combat aircraft over or from some countries, but they will not give permission. It is understandable that these countries may prefer not to be implicated because of the perceived risk of diplomatic backlash or should things in the region not go as the United States plans. This could cause problems in our increasingly expeditionary organization. UCAVs could help in these situations. With a stealthy UCAV as a self-contained strike package, the United States could fly missions over non-aligned countries without need for setting up air refueling (A/R) orbits to fuel up a strike package consisting of SEAD and strike aircraft and CSAR assets. The United States could simply advise a non-aligned nation that an attack would be transiting its airspace so that it knew it was not the target of the attack, as the United States has done in the past with cruise missile attacks. The figure 2 is a graphic representation of these two options.

---

[5]Department of Defense, *Unmanned Aerial Vehicles Roadmap 2000-2025,* A-13 - A-48.

[6]Bob Woodward, *Bush at War* (New York: Simon & Schuster, 2002), 152-167. According to Woodward, Gen Shelton and other senior military officials considered it a prerequisite to have rescue available before beginning combat air operations. In combat operations, the likelihood of successful recovery is highest when the aircrew is contacted for pick up in the first hour after going down.

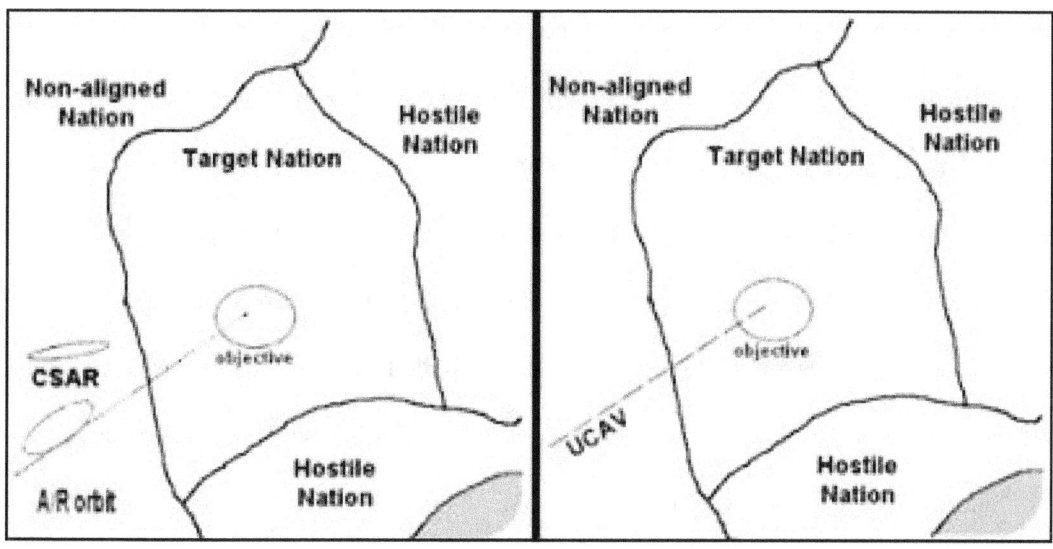

Figure 2. Two Possible Scenarios for Attacking a Target Nation

In the instance on the left, the nonaligned country can hardly hope to retain a neutral

status in the eyes of the target country, since it allowed a large aerial presence to use its airspace

for an attack on the target country. In the instance on the right, a stealthy UCAV may be able to

pull off the attack without giving any indication of its route of flight. Even if it were noticed, the

nonaligned country could still retain some hope of retaining neutral status in the eyes of the target

country by claiming ignorance of the presence of the UCAV(s) in its airspace.[7]

There are also diplomatic benefits that come from the nature of unmanned aircraft

themselves. The "U" in UCAV means there is no pilot at risk, during the flight or as a potential

prisoner of war, should the aircraft go down. As discussed in the chapter on the history of

UCAVs, evidence exists that when a manned aircraft is shot down over an adversarial country it

has strategic political and diplomatic implications at home and abroad (like the Francis Gary

Powers U-2 incident). When an unmanned aircraft is shot down, there is little political leverage

for the adversary to use (as in the Chinese shoot down of the BQM-34). Because of all of these

---

[7]"Pakistan Govt. Knew of U.S. Attack," Pakistan News Service, August 1998 [article on-line]; available from: http://paknews.com/aug98/main1aug-24 html; Internet; accessed on5 March 2003.

reasons, the UCAV will unmistakably be capable of serving the joint force commander in diplomatically sensitive situations, which is the sixth criterion.

Finally, the UCAV may be able to help attain one other important operational goal, that of theater missile defense. The desire to protect not only the homeland, but deployed troops was mentioned as the DoD's seventh new operational theme in chapter 4 of this paper, but it was not translated into a *required* joint team capability as listed in the Joint Staff white paper. At the very least, UCAVs will be useful in helping to spot these missile launches in the C4ISR role. But given the rapid attack, assess and re-attack capabilities of a UCAV with a HPM and advanced on-board radar they may be useful in helping to disable the missile in the boost phase as well. As a reminder, UCAVs equipped with the HPM will have been operationally fielded for three-to-five years by 2015. These UCAVs, properly integrated into a theater missile defense network with space- and ground-based early warning assets may provide the joint force commander a synergistic way to defeat the theater missile threat. The UCAV will be capable of at least augmenting in this role, and therefore meets the final criterion of utility for the joint force commander in 2015.

This chapter has used the information about what Air Force UCAVs will be capable of in 2015 and compared that to the seven criteria listed in the beginning of the monograph: rapid global projection; forcible entry into a joint operations area; immediate and synchronized employment of scaleable joint forces tailored to conduct simultaneously a broad range of joint military operations utilizing unified action; sustainment for extended periods in an austere environment or under anti-access or area denial conditions; and continuous joint C4ISR employment and knowledge generation fusion; the ability to employ a weapons system in diplomatically sensitive situations; and the capability to execute a portion of the theater missile defense mission. With each criteria, this chapter showed that the United States Air Force's UCAV of 2015 will meet or exceed it.

CHAPTER 6: CONCLUSION AND RECOMMENDATIONS

This monograph began with the fact that the Bush administration, senior members of congress, and the United States Air Force are all pushing for acquisition of UCAVs in significant numbers. The United States Air Force my be procuring UCAVs, but the wars of the United States are fought by joint force commanders. Because these UCAVs are going to end up augmenting or replacing other weapon systems in the joint combatant commander's range of force options, the United States military needs to know if they will be useful. Therefore, the research question of this monograph has been: Will uninhabited combat aerial vehicles have utility for the joint combatant commander in the year 2015? To answer this question, the author delineated seven capabilities that the UCAV must posses in order to provide utility to the joint force commander in 2015: rapid global projection; forcible entry into a joint operations area; immediate and synchronized employment of scaleable joint forces tailored to conduct simultaneously a broad range of joint military operations utilizing unified action; sustainment for extended periods in an austere environment or under antiaccess or area denial conditions; and continuous joint C4ISR employment and knowledge generation fusion; the ability to employ a weapons system in diplomatically sensitive situations; and the capability to execute a portion of the theater missile defense mission.

The history of UCAVs recounted in this paper has shown that while unmanned combat aircraft have been pursued since before World War I, only recently have UCAVs developed the broad range of capability and reliability required for combatant commanders to depend on them. This paper illustrated the range of missions of which the UCAV will be capable in 2015. UCAVs will be able to perform precision strike (and re-strike) using a range of kinetic and directed energy weapons, SEAD, and improved C4ISR integration while requiring either a small in-theater presence or no in-theater presence and very little human interaction. Additionally, UCAVs in

2015 will be able to either operate in concert with other combat aircraft as part of a combat package (SEAD, pre-strike reconnaissance, strike, and post-strike reconnaissance) or perform the roles of the whole package by itself. These UCAVs will be nearly autonomous and will handle long, boring and dangerous missions effectively while putting no onboard aircrew at risk.

The operating environment of the future calls for joint forces that are capable of meeting diverse challenges. To do this each force must be capable of: rapid global projection; forcible entry into a joint operations area; immediate and synchronized employment of scaleable joint forces tailored to conduct simultaneously a broad range of joint military operations utilizing unified action; sustainment for extended periods in an austere environment or under antiaccess or area denial conditions; and continuous joint C4ISR employment and knowledge generation fusion; allowing the commander to employ a weapons system in diplomatically sensitive situations; and executing a portion of the theater missile defense mission. The UCAV meets these requirements, answering the research question in the affirmative; uninhabited combat aerial vehicles will have utility for the joint combatant commander in the year 2015.

It is the author's opinion that UCAVs will not only be of utility to the joint force commander, but that they are likely to be heavily tasked in future operations. In researching this subject the author has become aware of the confluence of four figurative rivers which may drive interest in UCAVs even higher than it currently is. The first is that of the rapidly increasing capability due to synergy in technological advance in UCAVs and related fields, most of which is documented here. The second is that of the public (United States) expectation of fewer casualties in military operations. The third is the cost of replacing worn out manned weapon systems now being flown in their second year of combat operations in Operations Enduring Freedom and Iraqi Freedom; these aircraft will meet their life expectancy (in flying hours) much quicker than in peacetime. The fourth is the ongoing nature of the war on terror declared by President Bush on 20 September 2001 which may overstretch an already undermanned pool of pilots leading to retention problems.[1] The flow of this river may push the acquisition of UCAVs faster and in

greater number than the 14 X-45s programmed by 2006 and the 22 others for 2010-2012.[2]

Recommendations

Because of this predicted increase in interest in and dependence on UCAVs, the United

States military must be ready to employ these UCAVs fully. The first thing that should be done

to ensure that is education. Joint force commanders and planners should understand about their

increasing capabilities in addition to platform specific and general limitations. This can be done

easily with informational materials and briefings. While many people will recognize the benefits

of UCAVs in joint operations, there will remain some who do not know or believe in their

capabilities and may not readily accept UCAVs as a useful addition to the joint force. This can

be handled with demonstrations (on a range or at an air show) in order to build confidence.

The second thing that should be done to guarantee full utilization of UCAVs is their

integration into the common operational picture systems and tactical data links that will be

operational in the near future. Because the United States military must be ready to fight with a

joint team each time it fights, the UCAV must be fielded with the ability to integrate into the

systems which are going to be in place. This includes the informational systems of systems that

provide war fighters with real time positions, as well as the tactical data links which pass

targeting and threat information from one system to another on the battlefield. It appears to the

author that this is being done with systems that are planned for future fielding of communications

(like the ACN) in addition to the newly fielded system that passes target information to the

---

[1]President George W. Bush, "Address to a Joint Session of Congress and the American People," (speech delivered to a joint session of Congress, Washington, DC, 20 September 2001) [article on-line]; available from http://www.whitehouse.gov/news/releases/2001/09/20010920-8.html; Internet; accessed on 27 March 2003. Data on the pilot shortfall due to last through 2007: William W. Taylor, S. Craig Moore, and C. Robert Roll Jr., *The Air Force Pilot Shortage: A Crisis for Operational Units?* [document on-line]; available from http://www.rand.org/publications/MR/MR1204/; Internet; accessed on 27 March 2003, xi.

[2]Fulgham, 56-57.

AC-130 gunship. Even so, there are two reasons this may be of concern. The first reason is that some legacy systems with which UCAVs were not planned to integrate may remain into the future. The second reason is that because of the spiral development and quick acquisition, UCAVs may arrive on the scene before integrating (and deconflicting) systems will. This will be mitigated if the Air Force ensures that when UCAVs are "spiraled," the systems of record are either capable of accepting the UCAV or can be made to do so.

Another type of system is in use today with manned aircraft that may have to be changed in order to fully integrate UCAVs into air-to-ground missions. That is the close air support system. Currently, a trained tactical air controller using the radio must terminally control close air support missions. These missions could be done with UCAVs via a radio connection to the UCAV supervisor and either a laser designator or transmitted coordinates. Does the person on the ground in this instance need to be an Air Force trained person? Maybe some level of training should be performed, but the services may find that it is somewhat less than the current level. Even if this is not found to be the case, the services should prepare for including UCAVs into these missions so that the joint force commander can use the UCAV in any type of strike mission.

BIBLIOGRAPHY

Articles

Ackerman, Robert K. "Operation Enduring Freedom Redefines Warfare." *AFCEA International; The Fight for Freedom*," 22 March 2003. Article on line. Available from http://www.afcea.org/signal/tribute/EnduringFreedom.htm. Internet. Accessed on 5 March 2003.

Baker, Sue. "Global Hawk Program Enters Initial Acquisition." *Air Force News*, March 2001. Article on-line. Available from http://airforce.dtic.mil/news/Mar2001/n20010322_0402.shtml.

_____. "UAV Provide Warfighters a View." *Air Force News*, 12 October 2002, Article on-line. Available from http://www.af.mil/news/Dec2002/121002203.shtml. Internet. Accessed on 5 March 2003.

Belida, Alex. "Yemen al-Qaida Attack Appears to be Work of Unmanned CIA Plane," *SpaceDaily.com,* November 2002. Article on-line. Available from http://www.spacedaily.com/news/uav-02zm.html; Internet. Accessed on 5 March 2003.

Carmichael, Bruce W., Troy E. DeVine, Robert J. Kaufman, Patrick E. Pence, and Richard S. Wilcox. "Strikestar 2025." *Air Force 2025* Study. August 1996. Article on-line. Available from http://www.maxwell.af.mil/au/2025/volume3/chap13/vol3ch13.pdf. Internet. Accessed on 5 March 2003.

Callero, Monti D. "Assessment of Nonlethal Unmanned Aerial Vehicles for Integration with Combat Aviation Missions." Santa Monica, CA: Rand, 1995.

Fulgham, David A. "UCAV Spending Spikes in Pentagon Budget Plans," *Aviation Week and Space Technology*, 17 February 2003, 56-57.

NASA Dryden Flight Research Center. "Helios." 20 April 2002. Article on-line. Available from http://www.dfrc.nasa.gov/Research/Erast/helios.html. Internet. Accessed on 5 March 2003.

"Pakistan Govt. Knew of U.S. Attack." *Pakistan News Service,* August 1998. Article on-line. Available from: http://paknews.com/aug98/main1aug-24.html. Internet. Accessed on 5 March 2003.

"Roche Sorts it Out." *Air Force Association Magazine,* March 2002. Article on-line. Available from http://www.afa.org/magazine/March2002/0302roche_print.html. Internet. Accessed on 5 March 2003.

Rogers, Keith. "Reports Say Predators Seeing Action in War." *Las Vegas Review-Journal,* 20 October 2001, 24. Article on-line. Available from http://www.reviewjournal.com/lvrj_home/2001/Oct-20-Sat-2001/news/17263620.html. Internet. Accessed on 26 March 2003.

Smith, Michael. "Saddam to be Target of Britain's 'E-bomb'," 26 August 2002. Telegraph.co.uk on-line. Available from http://news.telegraph.co.uk/news/main.jhtml?xml=/news/2002/08/26/wirq26.xml. Internet. Accessed on 6 March 2003.

Taylor, William W., S. Craig Moore, and C. Robert Roll Jr. *The Air Force Pilot Shortage: A Crisis for Operational Units?* Document on-line. Available from http://www.rand.org/publications/MR/MR1204/. Internet. Accessed on 26 March 2003..

Tiron, Roxana. "Unmanned Bomber Prepares for Crucial Tests." *National Defense Magazine,* May 2002. Article on-line. Available from http://www.nationaldefensemagazine.org/article.cfm?Id=792. Internet. Accessed on 6 March 2003.

Tirpack, John A. "Send in the UCAVs." *Air Force Association Magazine*, August 2001. Article on-line. Available from http://www.afa.org/magazine/Aug2001/0801ucav.html. Internet. Accessed on 6 March 2003.

"UAV BATs A Few Direct Hits." Spacedaily.com, 23 October 2002. Article on-line. Available from http://www.spacedaily.com/news/uav-02zk.html. Internet. Accessed on 6 March 2003.

"UAV Makes Record-Breaking Transpacific Flight." Janes.com, April 2001. Article on-line. Available from http://www.janes.com/aerospace/military/news/misc/globalhawk010427_1_n.shtml. Internet. Accessed on 6 March 2003.

The Boeing Company. "Unmanned Combat Air Vehicle (X-45) Factsheet," 2003. Article on-line. Available from http://www.boeing.com/phantom/ucav.html. Internet. Accessed on 31 March 2003.

Wall, Robert, and David A. Fulgham. "New Tools Emerge For Info War Battle." *Aviation Week & Space Technology.* February 26, 2001. [On-line] Available from http://www.cadre.maxwell.af.mil/warfarestudies/iwac/Downloads/IW270%20Reading.doc. Internet. Accessed on 31 March 2003.

Wegerbauer, Cyndi. "Predator UAV Marks 50,000 Flight Hours." 30 October 2002. Article on-line. Available from http://www.ga.com/news/50000_flight.html. Internet. Accessed on 1 April 2003.

Books

Armitage, Michael, Sir. Air Power: Aircraft, Weapons Systems and Technology Series. Vol. 3. Unmanned Aircraft. London: Brassey's Defence Publishers, 1987.

Brooks, Rodney Allen . *Flesh and Machines: How Robots Will Change Us*. New York: Pantheon Books, 2002.

Martel, William C. *The Technological Arsenal: Emerging Defense Capabilities.* Washington, DC: Smithsonian Institute Press, 2001.

Reed, Arthur. *Brasssey's Unmanned Aircraft*. London: Brassey's; Elmsford, N.Y. : Distributed in the U.S.A. and Canada by the British Book Centre, 1979.

Woodward, Bob. *Bush at War.* New York: Simon & Schuster, 2002.

Zaloga, Steven J. "Reconnaissance Role Drives Growing Interest in UAVs." *1997-98 International Guide for Unmanned Vehicles.* AW&ST Yearbook. New York: Aviation Week & Space Technology; Washington, D.C.: Association for Unmanned Vehicle Systems International, 1997. Page 16.

Government Documents

Bush, George W. "Address to a Joint Session of Congress and the American People." Speech delivered to a joint session of Congress, Washington, DC, 20 September 2001. Speech on-line. Available from http://www.whitehouse.gov/news/releases/2001/09/20010920-8.html. Internet. Accessed on 5 March 2003.

Congressional Budget Office, "Options for Enhancing the Department of Defense's Unmanned Aerial Vehicle Programs," September 1998. Article on-line. Available from http://www.fas.org/man/congress/1998/cbo-uav.htm. Internet. Accessed on 5 March 2003.

DARPA. "Unmanned Combat Air Vehicle Summary Fact Sheet," April 2002. Article on-line. Available from http://www.darpa.mil/ucav/factSheet/factSheet.pdf. Internet. Accessed on 5 March 2003.

_____. " Unmanned Combat Air Vehicle Operational System," Defense Advanced Research Projects Agency Fact Sheet. Arlington, VA: DARPA, April 2002.

Glade, David. "Unmanned Aerial Vehicles: Implications for Military Operations." Occasional Paper No. 16, Air War College. Maxwell Air Force Base, AL: Air University Press, July 2000.

Kramlinger, George D. "Sustained Coercive Presence; Provide Comfort, Deny Flight, and the Future of Airpower in Peace Enforcement." Thesis, Air University Press, Maxwell Air Force Base, AL, February 2001. Article on-line. Available from http://www.maxwell. af.mil/au/aul/aupress/SAAS_Theses/Kramlinger/kramlinger.pdf. Internet. Accessed on 5 March 2003.

Longino, Dana A. "Role of Unmanned Aerial Vehicles in Future Armed Conflict Scenarios." Air University Research Report, Maxwell Air Force Base, AL.: Air University Press, 1994.

Rosenberg, Barry. "UAVs Step Into Combat Arena." Article on-line. Available from http://www.awgnet.com/shownews/01paris1/hardwr10.htm. Internet. Accessed on 6 March 2003.

Schwanhausser, Robert. "Unmanned Aerial Vehicles: From WWI Through Today." *1997-98 International Guide for Unmanned Vehicles.* New York: Aviation Week & Space Technology; Washington, D.C.: Association for Unmanned Vehicle Systems International, 1997. Page 10.

Secretary of the Air Force/Public Affairs. "UAV Technologies and Combat Operations," 1996. Document on-line. Available from http://fas.org/man/dod-101/sys/ac/docs/ucav96/chap3. pdf. Internet. Accessed on 26 March 2003.

US Central Intelligence Agency. US National Intelligence Council. *Global Trends 2015: A Dialogue About the Future With Nongovernment Experts.* Washington, DC: Government Printing Office, 2002.

US Department of Defense. *DOD Dictionary of Military and Associated Terms.* Washington, DC: Government Printing Office, 14 August 2002. Document on-line. Available from http://www.dtic.mil/ doctrine/jel/doddict/data/i/index.html. Internet. Accessed on 6 March 2003.

_____. *Quadrennial Defense Review Report, September 30, 2001.* Washington, DC: Government Printing Office, 2001.

_____. *Unmanned Aerial Vehicles Roadmap 2000-2025.* Washington, DC: Government Printing Office, April 2001.

US Joint Staff. "An Evolving Joint Perspective: US Joint Warfare and Crisis Resolution in the 21st Century." White Paper. Washington, DC: Government Printing Office, January 2003.

Walker, Jan. "UCAV Named Best of What's New, Completes Low Speed Taxi Tests." Defense Advanced Research Projects Agency News Release, 4 December 2001. Article on-line. Available from http://www.darpa.mil/body/NewsItems/wordfiles/UCAVfinal.doc. Internet. Accessed on 3 March 2003.

Walling, Eileen M. "High Power Microwaves; Strategic and Operational Implications for Warfare." Occasional Paper No. 11, Air War College. Maxwell AFB, AL: Air University Press, May 2000. [On-line] Available from www.au.af.mil/au/awc/awcgate/cst/csat11.pdf. Internet. Accessed on 31 March 2003.